Sidney Barwise

Practical Hints on the Analysis of Water and Sewage

For the use of medical Officers of Health, medical Practitioners, sanitary

Engineers, etc

Sidney Barwise

Practical Hints on the Analysis of Water and Sewage
For the use of medical Officers of Health, medical Practitioners, sanitary Engineers, etc

ISBN/EAN: 9783337139551

Printed in Europe, USA, Canada, Australia, Japan

Cover: Foto ©ninafisch / pixelio.de

More available books at **www.hansebooks.com**

PRACTICAL HINTS ON THE ANALYSIS OF WATER AND SEWAGE

FOR THE USE OF

MEDICAL OFFICERS OF HEALTH, MEDICAL PRACTITIONERS, SANITARY ENGINEERS, ETC.

BY

SIDNEY BARWISE, M.D. (Lond.), D.P.H. (Camb.),

Fellow of the Sanitary Institute, Fellow of the Incorporated Society of
Medical Officers of Health, Medical Officer of Health Derbyshire
County Council, Author of " The Purification of Sewage," etc.

PRICE TWO SHILLINGS AND SIXPENCE.

London:

REBMAN, LIMITED,

129, SHAFTESBURY AVENUE, CAMBRIDGE CIRCUS, W C.

1899.

CONTENTS.

CHAPTER I.

Introduction.

CHAPTER II.

The Apparatus required for Water Analysis.

CHAPTER III.

The Standard Solutions used in Water Analysis.

CHAPTER IV.

The Methods of Water Analysis.

Contents

CHAPTER V.

The Interpretation of the Results of Analysis.

CHAPTER VI.

On the Bacterioscopic Examination of Drinking Waters.

TABLES.

PRACTICAL HINTS ON THE ANALYSIS OF WATER AND SEWAGE

FOR THE USE OF MEDICAL OFFICERS OF HEALTH.

CHAPTER I.

Introduction.

ERRATUM.

Page 33, paragraph 28, line 4, *for* " 10 c.c. of Nitrate of Potash,"
read " 1 c.c. of Nitrate of Potash."

actual experience in practical sanitation, as he will have to decide what may be considered a safe distance from sources of pollution under different geological and structural conditions.

A point which has to be borne in mind is that there is no fixed distance which may be called a *safe* limit from sources of pollution ; each case must be judged upon its merits. For, with close sandy rock in which the subsoil water travels through the pores of the rock, purification by bacterial action and subsidence may be complete in a comparatively short distance ; on the other hand, with a very pervious rock lying on a bed of clay, with flagstones, shales,

Contents

CHAPTER V.

The Interpretation of the Results of Analysis.

CHAPTER VI.

The Bacterioscopic Examination of Drinking Waters.

PRACTICAL HINTS ON THE ANALYSIS OF WATER AND SEWAGE

FOR THE USE OF MEDICAL OFFICERS OF HEALTH.

CHAPTER I.

Introduction.

THE question " What is a pure and wholesome water ?" is one which, in addition to its chemical aspect, involves a variety of considerations, and requires, before it can be answered, several distinct kinds of knowledge—knowledge which, as a rule, has only been acquired by a medical man specially trained in sanitary science. Such points as, what quantity of zinc or lead in a water, or what amount of sulphate of calcium, will prove injurious to health, are considerations in practical medicine demanding a knowledge of therapeutics and physiology. A thorough familiarity with the chemistry of water analysis is also necessary, and for the correct interpretation of the results of analysis the medical officer must be well versed in all such general hygienic considerations as the circulation of underground water in the various geological formations and the changes it thereby undergoes. A knowledge of microscopy and bacteriology is also necessary. In addition, a careful inspection of the source of the water should always be made, and this inspection, in the case of public supplies, should only be entrusted to one who has had actual experience in practical sanitation, as he will have to decide what may be considered a safe distance from sources of pollution under different geological and structural conditions.

A point which has to be borne in mind is that there is no fixed distance which may be called a *safe* limit from sources of pollution ; each case must be judged upon its merits. For, with close sandy rock in which the subsoil water travels through the pores of the rock, purification by bacterial action and subsidence may be complete in a comparatively short distance ; on the other hand, with a very pervious rock lying on a bed of clay, with flagstones, shales, mountain limestone, and other formations, the subsoil water travels in well-defined cracks and channels, and little or no purification can be effected. The outbreak of typhoid fever at New Herrington (Durham) was shown by an inspector of the Local Government Board to have been due to the pollution of subsoil water from a sewage farm three-quarters of a mile away. So also at Lausen, in Switzerland, sewage matter, after percolating through gravel for the distance of a mile, caused an outbreak of typhoid fever by gaining access to a public water-supply.

The all-important point to be ascertained is the direction of the flow of the subsoil water; this may be in the direction of, or in the reverse direction to, the fall of the surface. It is in the direction of the dip of the strata, but below the level of the plane saturation, it may even be in the opposite direction to the dip of the strata.

Bearing in mind the experience gained from the Maidstone epidemic of enteric fever, it should hardly be necessary to lay stress upon the need of a complete inspection of the *source* of a water. The Local Government Board has done excellent service in always insisting upon this part of the inquiry. Indeed, not infrequently the Board employs a practical geologist to go over the district from which a water-supply is derived in company with the medical inspector, the results of the analysis of the water being interpreted in the light of the information obtained by the inspection of its source and the investigation of its pedigree. Unless a careful inspection of the source of a water be made, calamitous results may at any time follow, as in the Tees Valley outbreak of enteric fever.

In this instance several distinguished chemists pronounced a water to be " pure and wholesome " which was found by Dr. Barry, of the Local Government Board, to be receiving, before it was filtered, crude sewage and the soakage from privy-middens. Sir E. Frankland, F.R.S., reported the water in question to be "free from any trace of sewage contamination"; Mr. A. H. Allen, F.I.C., said his results "negatived any suspicion of contamination by sewage or cesspool drainage"; Messrs. Patterson and Stead declared the water to be "free from any sewage contamination"; and Mr. F. K. Stock, the County Analyst, declared it to be "good and wholesome drinking-water." Yet an inspection of the gathering-ground proved that sewage and the soakage from cesspools drained freely into this water, of which Sir Richard Thorne Thorne subsequently wrote: " Seldom, if ever, has the proof of the relation of the use of water so befouled to wholesale occurrence of typhoid fever been more obvious or patent."

As another instance bearing on this point I quote the following from Dr. Thresh : " Within the last few weeks I have detected a cesspool within twenty yards of a collecting well of a public water company, the cesspool receiving the sewage from a group of cottages. I had a short time previously examined the water, and found it satisfactory from both a chemical and bacteriological point of view. After the cesspool was discovered I took samples of the water. The water flowing in on the side nearest the cesspool contained more bacteria and more free ammonia than the water entering on the opposite side; neither contained any appreciable amount of organic matter."

Several similar cases are recorded by the Medical Department of the Local Government Board. It is therefore, in the opinion of the author, desirable that *the results of the analysis of a water should be interpreted by the medical officer who has inspected its source*, or, possibly, that the analysis should be undertaken by the medical officer of health himself, who in the course of his ordinary duties will be called upon to inspect the surroundings of every well and source of water in his district which is suspected of having given rise to any outbreak of disease.

Again, every medical officer of health of a rural district will have in the area under his supervision many wells the quality of the water in which he will from time to time have occasion to inquire into promptly. To send the water to an analytical chemist may mean delaying until the sanitary authority meets, and even then it is frequently useless to recommend to the authority that a water should be sent for analysis, as this course entails some little expense. On the other hand, if the medical officer is himself in a position to analyze the water, and finds upon analysis results which confirm the suspicions aroused by an examination of its source, he is in a position to act much more effectively and promptly than if he were dividing his responsibility.

Before any house can be occupied in a rurul district, the medical officer of health, or surveyor, must report to the authority that there is within a reasonable distance an available supply of wholesome water (Public Health (Water) Act, 1878, Section 6). In order to carry out this duty it is necessary that the sources of the supply should be thoroughly investigated. If it be a surface well, the direction of the flow of the subsoil water should be ascertained. It should be seen that the ground above is free from possible sources of occasional pollution. Any drains near must be tested, to see that they are water-tight, and the well itself must be examined, to see that it is properly constructed to keep out surface pollutions. The dry-weather yield of the source of supply must be proved to be sufficient, and the quality of the water must be ascertained by analysis, bearing in mind that the water from a new well will at first necessarily contain a considerable amount of organic matter.

On this head I cannot do better than give the words of Dr. Thresh, one of the highest living authorities on water-supplies : " It is far more important to examine the source of a water than to analyze it . . . the most carefully made analyses may be seriously misleading if interpreted without the knowledge obtained from such examination."

The following pages are designed for the use of medical officers of health, whether in private practice or devoting their whole time to the duties of their office. They will also be found useful by medical men studying for a diploma in public health.

A large number of labour-saving devices have been taken advantage of, and the apparatus has been specially designed to take up little room, so that an accurate and complete hygienic examination of a water can be made by any medical officer if he has at his disposal gas and a supply of good water, the whole of the apparatus when in use only occupying a table space of about 3 feet by 2.

Messrs. Southall Bros. and Barclay Ltd., of Birmingham, have spared no pains in the production of the necessary apparatus in a convenient form, and at a moderate price.

CHAPTER II.

The Apparatus required for Water Analysis.

A GOOD BALANCE WITH WEIGHTS ON THE METRIC SYSTEM.

1. A suitable balance is Becker's, No. 28*a*, which will be found extremely useful for all purposes. The balance is fitted into a glass case with a sliding door. It is absolutely necessary to weigh the solid matter left upon evaporating a water to dryness rapidly, and in a closed balance, as this solid matter is frequently extremely deliquescent. On this account it is advisable to keep in the balance case a small beaker about a quarter full of strong sulphuric acid in

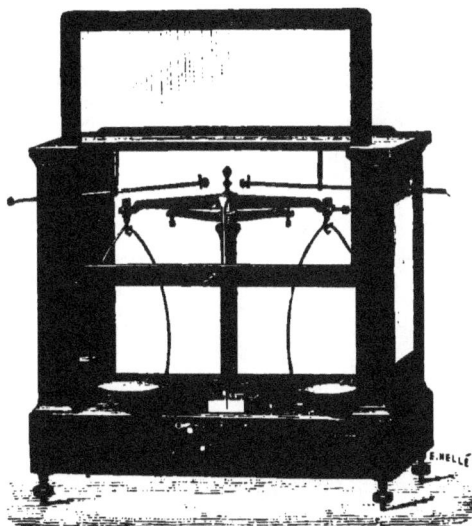

FIG. 1.—THE BALANCE.

which a piece of pumice-stone is floating; the acid absorbing the moisture from the air, thereby conduces to much greater accuracy in weighing. Instead of sulphuric acid a few small pieces of freshly-burnt quicklime may be used; but this will require renewing more frequently than the acid.

The balance must be fixed on a piece of plate-glass or slate, in a firm position, where it will be free from vibration and rapid changes of temperature. By means of the screws it must be set in a horizontal position so that the indicator needle stands at zero. If after thoroughly dusting and levelling the balance the pointer does not stop at 0, it can be brought there by carefully adjusting the small brass nut on the end of the right-hand limb of the balance beam.

Before weighing anything it is necessary to see that the pans are free from dust. The central screw should be turned to the left so as to release the beam, and the little knob to the left of the central screw must be gently pressed to release the two pans, then if the needle points to zero the balance is ready for weighing.

THE WEIGHTS.

2. The weights, with a pair of brass forceps for lifting them, are contained in a small box, and they have the following values :

Brass :

	50 grammes.	20 grammes.	10 grammes.	10 grammes.
	5　　,,	2　　,,	1 gramme.	1 gramme.

Platinum :

·5 gramme	·2 gramme	·1 gramme	·1 gramme
(500 milligrammes).	(200 milligrammes).	(100 milligrammes).	(100 milligrammes).
·05 gramme	·02 gramme	·01 gramme	·01 gramme
(50 milligrammes).	(20 milligrammes).	(10 milligrammes).	(10 milligrammes).

They are arranged in the box as follows, and should always be immediately replaced in their proper places in the box when not in use on the pan of the balance.

50	20	10	10
5	2	1	1
·5	·2	·1	·1
·05	·02	·01	·01

FIG. 2.—THE WEIGHTS.

The milligrammes themselves (*i.e.*, ·001 to ·009 gramme) are determined by the position of the rider on the right-hand beam of the balance.

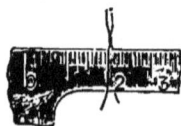

FIG. 3.—THE RIDER.

As soon as the correct weight is ascertained, it should be entered at once in a small notebook which should be kept in the drawer of the balance with the weights, the weighing being again checked by the removal of the weights from the pan of the balance to their proper places in the weight-box.

The Dish for the Estimation of the Total Solids.

3. The use of a nickel dish is recommended for the estimation of the total solids, the only drawback to the use of nickel being that it cannot be heated over a Bunsen flame as a platinum dish can, and greater care must be exercised in cleaning it. To clean, it should be soaked in a 1 per cent. solution of hydrochloric acid for a few minutes, and then be rubbed bright with the finest Calais sand. The acid must not be stronger than 1 per cent., or it will attack the

Fig. 4.—Improvised Water-bath.

nickel. The dish should be rinsed out with clean water, dried with a clean cloth, and heated to 100° C. in a steam-oven, or on a water-bath. The dish should then be rapidly wiped and placed in the desiccator to cool over sulphuric acid.

When cool, it is to be placed on the left-hand pan of the balance, and weighed as rapidly as possible, and the tare weight entered at once in the weighing-book.

It will be found that after each cleansing the dish gets one or two milligrammes lighter, but with careful use the loss should not be greater than this.

After estimating the total solids in a sample of water, the dish should be cleaned, dried, and weighed, and kept in the desiccator ready for use.

The Water-bath.

4. The solid matter in a water is determined by evaporating a known quantity of water to dryness over a water-bath. An ordinary beaker will serve as a water-bath, as shown in Fig. 4, the nickel dish being squeezed slightly to permit the escape of steam from under it.

One of the handiest arrangements for the estimation of total solids is a combined water-bath and steam-oven. This water-bath has holes for two dishes. If it is desired to use the apparatus for only one dish, the other hole may be filled with an inverted bottle

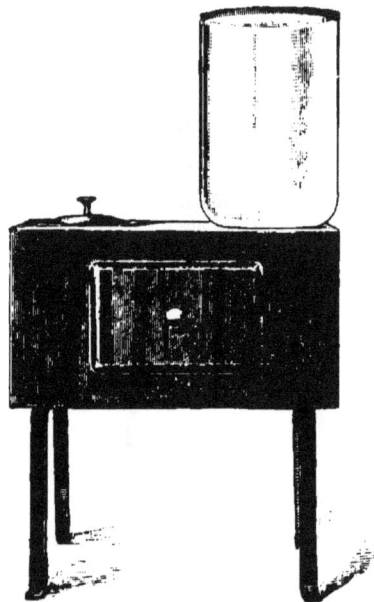

Fig. 5.—Combined Water-bath and Steam-oven.

of water, which will replenish the bath as the water is evaporated away. The bottle should be corked by a perforated cork, pierced by a piece of glass tubing about the diameter of a lead pencil, and projecting to about the level of the middle of the water-bath. See Fig. 5.

The Steam-oven.

5. After the water in the nickel dish has been evaporated down to dryness, the dish should be wiped thoroughly clean on the outside, and then dried for half an hour in a water-oven. The apparatus described above, and illustrated in Fig. 5, is provided with a small drawer. This drawer is made to accommodate a 100 c.c. nickel dish. It is heated at the same time as the bath, and can be used as an oven, thereby saving heat and time. When the solid matter in the nickel dish has been thoroughly dried in the steam-oven, the

FIG. 6.—THE STEAM-OVEN.

drawer is taken out by means of a pair of crucible tongs, and carried on a duster to the desiccator, the nickel dish being at once removed, while hot, to the desiccator.

Unless such a bath is used, it will be necessary to be provided with a separate steam-oven, such as shown in Fig. 6.

THE DESICCATOR.

6. As many of the salts left on evaporating a water to dryness are of a highly hygroscopic character, the nickel dish should, on removal from the steam-oven, be at once put to cool in an air-tight desiccator. This is a glass vessel constricted at its middle, the bottom compartment containing strong sulphuric acid, and several pieces of pumice-stone. The nickel dish rests on a pipe-clay

FIG. 7.—DESICCATOR.

triangle, placed upon a piece of perforated zinc, which forms the bottom of the upper chamber of the desiccator as shown in Fig. 7. The lid of the desiccator must be kept greased with vaseline.

The desiccator should be kept on the left of the balance, and the drawer containing the nickel dish should be carried from the steam-bath to the desiccator, so that the dish may be placed in the desiccator while still hot.

THE CONDENSER FOR DISTILLATION.

7. As a large quantity of ammonia-free distilled water is indispensable for the general purposes of analysis, an apparatus for making it is advisable, although such water can be purchased, in

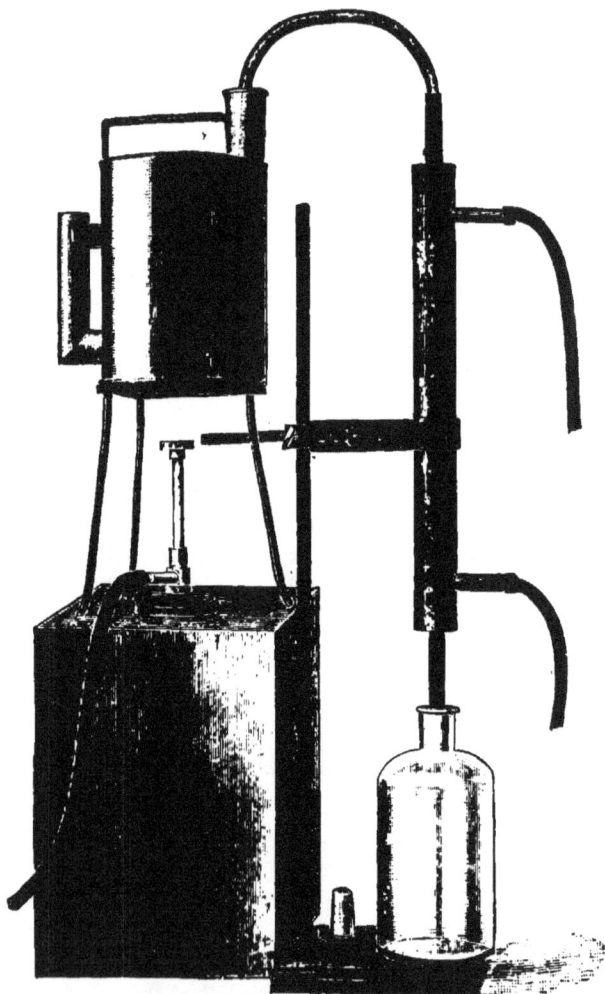

FIG. 8.—APPARATUS FOR PREPARATION OF DISTILLED WATER.

glass-stoppered Winchester quart bottles, from the leading chemists. It is possible to use the same condenser as is employed in the process of estimating the free and organic ammonia in a water, but if many sewage effluents or a large number of waters are to be analyzed, another condenser should be obtained.

The condenser devised by the author occupies very little space, and consists of a block-tin pipe, soldered within a copper pipe, water being circulated up between the two pipes.   To make

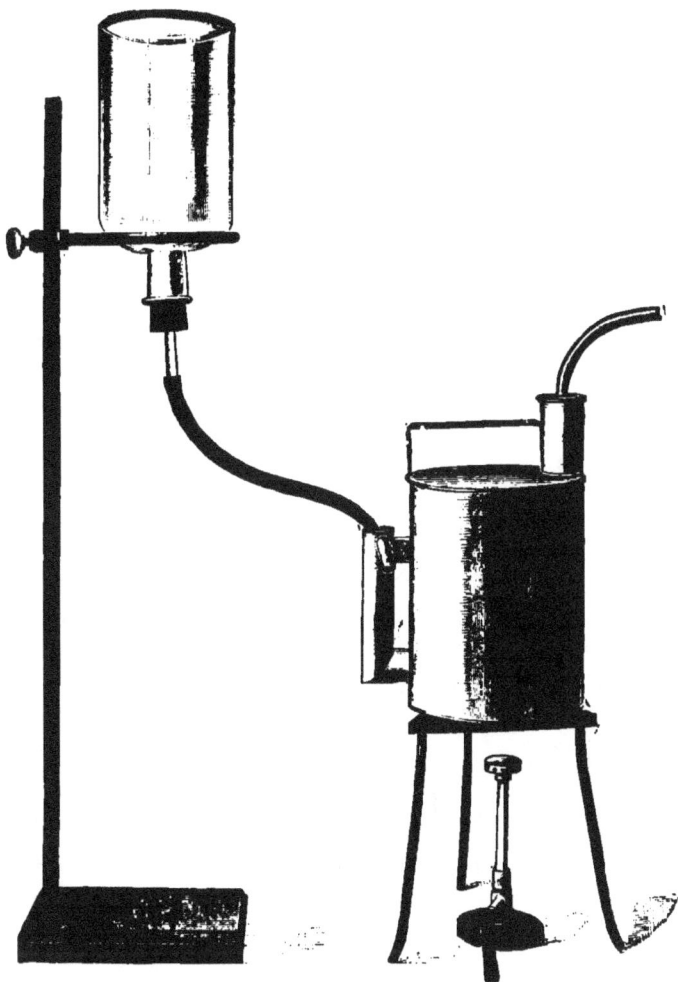

FIG. 9.—AUTOMATIC FEED OF DISTILLING APPARATUS.

ammonia-free distilled water, a tin can is fitted with a semi-circular block-tin pipe, and is filled with 1,500 c.c. of good tap or spring water. A cork is fitted into the filling hole, and the tin pipe is fixed within the tin pipe of the condenser. The can is then rested on a tripod, with a Bunsen burner underneath, and a stream of water passed upwards through the condenser. The distilled water is caught, and to successive 50 c.c. of it 2 c.c. of Nessler's test are added until no discoloration takes place, showing that the water is ammonia-free. With good tap waters this will be the case when 200 c.c. have been distilled off. The distillate is to be rejected until it shows absolutely no trace of ammonia. The distillation should be continued until a litre (1,000 c.c.) of ammonia-free water has been collected. In any case the distillation should not be continued after 1,200 c.c. of the original 1,500 c.c. have been distilled off, or the still may be injured.

The arrangement described is shown in Fig. 8.

Where a good public water-supply is available, it will frequently happen that the ammonia-free water may be obtained even in the first portion of the distillate. When this is the case, water may be supplied to the tin can continuously to replace the water distilled over, either by allowing water to drop into the hole at which the tin can is filled, or by the same arrangement by which it is suggested that the water-bath should be filled, namely, by inverting a bottle of water over the filling hole.

As soon as the water in the still falls below the level of the bottom of the glass tube through the cork, a bubble of air will be admitted into the bottle, and a small quantity of water will run into the still.

Before using the condenser for the preparation of ammonia-free distilled water, it should be washed out with good tap water, and after use the ends should be plugged with corks or cotton-wool to keep out the dust. The tin can should be treated in the same manner.

The Apparatus for the Albuminoid or Organic Ammonia Process.

8. The old way of conducting this process was extremely troublesome, but with the aid of the apparatus here described it can be completed in about half an hour. The use of a retort is dispensed with ; it is no longer necessary to remove the flame at the end of the first part of the process, and consequently the risk of breakage is reduced to a minimum. In addition to this, the apparatus is exceedingly cheap, the only part which is breakable costing one shilling.

The condenser used is of the same pattern as that employed for making the distilled water. Instead of a retort in which to distil

the water to be analyzed, a *wide-necked* Florence flask is used, hold-
ing about 40 ounces. This is fitted with a good doubly-perforated
indiarubber bung. Before use the bung is boiled in a solution of
caustic soda, to rid it of all traces of ammonia.

Through one perforation is fitted a piece of curved block-tin
piping, the other end of which is made to taper so as to fit into the

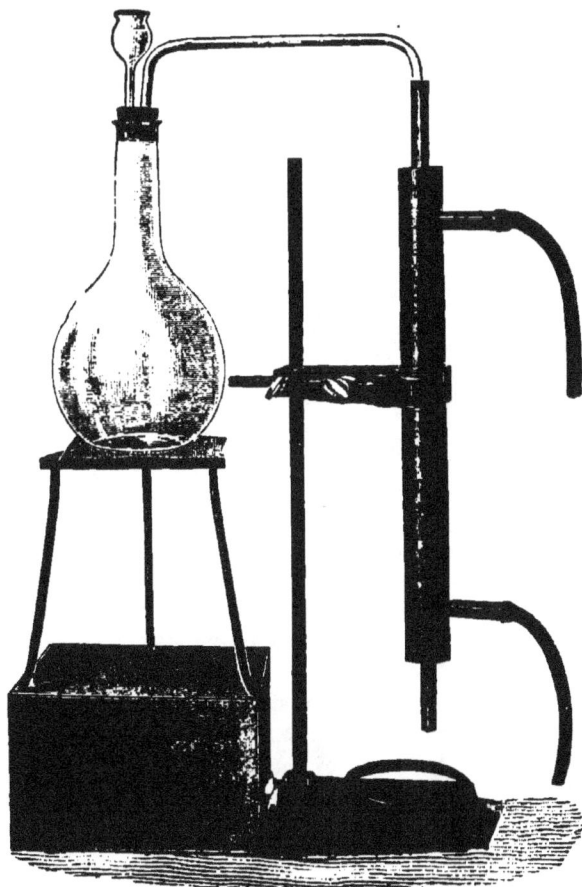

FIG. 10.—APPARATUS FOR AMMONIA PROCESS.

tin pipe of the condenser. The second perforation is intended for
a thistle funnel, into which is fitted a tapering glass rod.* The
thistle funnel, as will be explained later, permits the alkaline
permanganate solution to be poured into the flask after the free
ammonia is distilled off, without it being necessary to turn down
the flame, or to remove the bung in the flask. The end of the

* Fig. 10 does not show the glass rod plugging thistle funnel.

thistle funnel should be partly sealed to make the permanganate run in a small, steady stream.

It will be well to be provided with three distilling flasks.

9. Half a dozen 100 c.c. Nessler glasses are also necessary. It will be found a great convenience to have one of these graduated in five parts, and tenths of a part, to be used for the standard solution of ammonia. Four of the Nessler glasses should be numbered consecutively, and risk of error will be reduced to a minimum by using them on the Nesslerizing stand devised by the author. The Nessler glasses 1 and 2 are for collecting the free ammonia; numbers 3 and 4 are for collecting the organic ammonia; the graduated tube is for making the standard solution, and the re-

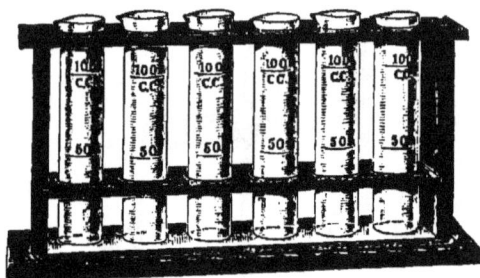

Fig. 11.—Nessler Tube Stand.

maining tube is for making an additional tube of standard solution, or for collecting more free ammonia if it does not all come off in the first 200 c.c. of distillate.

Fig. 11 illustrates the Nessler stand recommended; the tubes rest in tapering holes.* The bottom of the stand is covered with opal glass, so that the depth of the colour may be easily seen by looking down into it through the cylinder. When the tubes are not in use they should be swilled out with clean water and kept inverted on the stand.

10. A large (2 litre) porcelain or enamelled iron dish is required for making the alkaline permanganate solution used for the organic ammonia process.

11. A measuring glass with a lip, and graduated to 50 c.c., is required to pour the 50 c.c. of the alkaline permanganate through ·

* In the illustration the tubes are shown resting on the bottom of the stand instead of in the tapering holes above the opal glass.

the thistle funnel into the distillation flask after all the free
ammonia has been given off, in order to disengage the organic
ammonia. This measure should have a firm bottom, and pour
steadily (Fig. 12).

A half-litre (500 c.c.) measuring flask (such as is shown in Fig. 13)
is required for measuring the quantity of water wanted for distillation in estimating the free and organic ammonia.

A 100 c.c. flask is wanted for measuring the quantity of water for
evaporation in the nickel dish for estimating the total solids, and

FIG. 12.   FIG. 13.   FIG. 14.

for the estimation of chlorides. A 50 c.c. graduated cylinder will
be convenient for measuring the water required for the estimation
of hardness (Fig. 14).

By always taking the same quantity in the same flask for the
various processes, the method of analysis becomes automatic, and
accuracy is insured.

BURETTES AND A BURETTE-STAND.

12. Two burettes, one for standard nitrate of silver for estimating
the chlorine, and one for soap solution for the estimation of hardness, are also necessary. The burettes should each be labelled, so
that the same burette is always used for the same purpose. When
they have been finished with, the remaining standard solutions may
be run off into the stock-bottles. The burettes should then be
drained dry, plugged with cotton-wool to keep out dust, inverted
on the burette-stand, and kept in this position with the taps open.
Before use a small quantity of the appropriate standard solution
should be run through each tap, and before each estimation the
burettes should be filled, so that the bottom of the curved surfaces
of the standard solutions stand accurately at 0.

A metal burette-stand is recommended, as it is firmer than a
wooden one.

A 200 c.c. porcelain dish is required for the estimation (by the nitrate of silver solution) of the chlorine as chlorides.

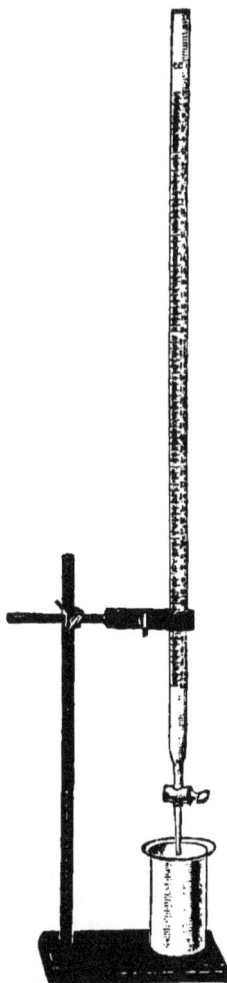

Fig. 15.—Metal Burette-stand and Burette.

PIPETTES.

13. A 5 c.c. graduated pipette, as shown in Fig. 16, for measuring the standard solution of ammonia is used, and should be kept as far as possible for this use.

A pipette to deliver 10 c.c. is used to measure the quantity of water required for the estimation of nitrates, and also in the oxygen-absorbing process, and in the former process a 1 c.c. pipette to deliver 1 c.c. is also used to measure the phenol-sulphuric acid.

FIG. 16.    FIG. 17.

THREE SMALL BEAKERS.

14. For the estimation of the nitrogen as nitrates, three small 25 c.c. beakers are necessary for evaporating to dryness on the water-bath the 10 c.c. of water used in this process. The beakers should be scratched 1, 2, and 3 with a diamond, or be marked — ⌶ ☰ with a file, so that no mistake can arise when two beakers and a standard are being evaporated to dryness at the same time.

In addition to the above appliances, which are specially designed for the use of medical officers of health, the following apparatus is also necessary:

Two funnels are required—a large (6-inch) funnel for filling the distillation flask for the ammonia process, and a small (3-inch) funnel for making the standard solutions.

Two watch-glasses—one large and one small.

Two of Fletcher's safety Bunsen burners.

One iron tripod.

A pair of crucible tongs.

A packet of 3-inch Swedish filter-papers.

Several lengths of glass tubing.

A few pieces of glass rod.

A couple of white glass 12-ounce stoppered bottles for estimating the oxygen absorbed.

A couple of white glass 8-ounce stoppered bottles for the estimation of the hardness.

CHAPTER III.

The Standard Solutions used in Water Analysis.

15. All the standard solutions required may be purchased ready for use from the leading chemists. Of course it is desirable to know how to make these solutions, and with the apparatus described in the previous chapter they may be prepared quite accurately. Ammonia-free water is an essential for this purpose. If a good public water-supply which easily yields ammonia-free water is not obtainable, it will be necessary to purchase ammonia-free water, and this should always be at once tested with Nessler's test (sec. 16, *b*). If it gives the slightest discoloration it should be returned.

To make a litre of a standard solution, measure half a litre of distilled water in the 500 c.c. measuring-flask, the same having been first thoroughly cleansed and rinsed out with distilled water. Pour the water carefully through the 3-inch funnel into the bottle in which the standard solution is to be kept, the bottle having also been previously thoroughly cleansed and rinsed out with distilled water, and fill the measuring-flask again with distilled water. The quantity of the reagent required to make the standard solution is then accurately weighed on a watch-glass and placed within the funnel whilst in the neck of the bottle. The reagent is then to be forced into the bottle through the funnel by means of a clean glass rod, and washed down with the second 500 c.c. of distilled water. Of course it is absolutely necessary that not only should none of the reagent be lost, but also that none of the water should be spilt. In the first instance it will be wise to buy a small stock of accurate standard solutions so as to check the accuracy of the balance, the other apparatus, and the method of working.

16. The following is a list of the standard solutions required. All the reagents must be purchased as chemically pure, and be kept in a dark cupboard in well-stoppered bottles.

FOR THE DETERMINATION OF THE FREE AND ORGANIC AMMONIA.

(*a*) *Standard Solution of Ammonia.*

Dissolve 3·150 grammes of ammonium chloride in 1 litre of water. Label: STRONG AMMONIUM CHLORIDE SOLUTION.

Pipette off 5 c.c. of this strong solution (when it is completely dissolved and properly mixed) into the 500 c.c. measuring-flask, and make up to 500 c.c. with ammonia-free water. Label: Standard Ammonia Solution (1 c.c. = ·00001 gramme of ammonia).

(b) Nessler's Solution.

. Dissolve 62·5 grammes of iodide of potash in 250 c.c. of distilled water ; put 1 c.c. of this solution aside. Make a *cold saturated solution* of corrosive sublimate by dissolving 35 grammes of powdered perchloride of mercury in 600 c.c. of warm distilled water. When this is cold add it gradually to the iodide solution until a slight precipitate appears. The precipitate should then be redissolved by adding as little as possible of the 1 c.c. of iodide of potash solution set aside. Add 150 grammes of caustic potash dissolved in 150 c.c. of water. This will make about a litre of Nessler's test. Set aside in a well-stoppered bottle and decant off the clear portion for use. Label : Nessler's Test.

(c) Alkaline Permanganate Solution for Organic (Albuminoid) Ammonia.

Dissolve in the large porcelain dish 7 ounces of caustic potash sticks in 1,100 c.c. of ammonia-free distilled water ; add 8 grammes of pure permanganate of potash and boil down to 1,000 c.c. If 50 c.c. distilled with 250 c.c. of ammonia-free water yield *no* ammonia, the solution is accurately prepared. If it yields ammonia it requires further boiling to drive off the ammonia. Care should be exercised that no dirt off the caustic potash bottle falls into the dish. The best way to measure the 7 ounces is to counterpoise the big dish in a good pair of scales, and then take out the sticks of potash with a clean pair of forceps which have been heated in a Bunsen burner, transferring the sticks of potash to the dish until 7 ounces is turned. Label : Alkaline Permanganate Solution.*

17. For Chlorine.

(a) Standard Nitrate of Silver Solution.

Dissolve 4·788 grammes of nitrate of silver in 1,000 c.c. of distilled water. Label: Standard Nitrate of Silver Solution (1 c.c. = ·001 gramme of chlorine. With 100 c.c. of water, each c.c. = 1 part per 100,000).

(b) Indicator of Potassium Monochromate.

Dissolve 1 part of the pure salt in 10 of water, and keep in a drop-bottle.

---

* The caustic potash is sold in 1-pound bottles. The remaining 9 ounces should be at once transferred to a large wide-mouthed glass-stoppered bottle, as it does not keep in the corked bottles after they have once been opened.

18. For Nitrates.

(a) *Standard Potassium Nitrate Solution.*

Dissolve 0·722 gramme of recently fused potassium nitrate in 1,000 c.c. of water. Label: Standard Nitrate Solution (10 c.c. = ·001 gramme of nitrogen).

(b) *Phenol Sulphuric Acid.*

Add 6 grammes of pure phenol and 3 c.c. of distilled water to 37 c.c. of pure sulphuric acid free from nitrates and digest for several hours at about 180 F.

(c) *Liquor Potassœ.*

Make a 5 per cent. solution of caustic potash (free from nitrates) in distilled water, or use liquor potassæ B.P.

19. For Hardness.

Standard soap solution is largely used in commerce, and is prepared by all the leading chemical houses. As it is troublesome to make, and does not keep well, it is better to obtain it ready made. The method of making the soap solution is fully described in the British Pharmacopœia.

20. For the Oxygen Consumed Process.

(a) *Solution of Potassium Permanganate.*

Dissolve 0·895 gramme of potassium permanganate in 1 litre of freshly distilled water. Label: Permanganate Solution (10 c.c. = ·001 gramme of oxygen).

(b) *Sodium Thiosulphate Solution.*

Dissolve 1 gramme of pure crystallized thiosulphate in 1 litre of distilled water.

(c) *Dilute Sulphuric Acid 1 in 4.*

Add gradually 100 c.c. of pure sulphuric acid to 800 c.c. of distilled water.

(d) *Potassium Iodide.*

Dissolve 1 part of the pure recrystallized salt in 10 parts of distilled water.

(e) *Starch Solution.*

Take about a gramme of rice starch and work into a cream with cold water. Pour into 100 c.c. of boiling water and continue boiling for two minutes. When cold, decant off clear liquid for use. The solution does not keep, and must be freshly prepared.

21. STANDARD SOLUTIONS FOR THE ESTIMATION OF LEAD.

(a) *Standard Lead Solution.*

Dissolve 0·183 gramme of crystallized acetate of lead in 1 litre of water acidulated with acetic acid. Label : STANDARD LEAD SOLUTION (1 c.c. = ·0001 gramme of lead).

(b) *Sulphide of Ammonium Solution.*

22. The following reagents are also necessary :

1. Hydrochloric acid for cleaning distillation flasks, etc.
2. Pure sulphuric acid.
3. Solution of sulphuretted hydrogen.
4. Methyl orange solution 1 per cent., as an indicator for acids other than carbonic.

CHAPTER IV.·

The Methods of Water Analysis.

23. In the first place it should be clearly understood that every dish, tube, and flask must be absolutely clean before use. This is a simple matter when there is a good public water-supply; where, however, the public supply is of indifferent quality, it will be necessary to rinse everything out with distilled water after it has been washed with tap water.

The water for analysis should be collected in a glass-stoppered Winchester quart bottle; but if care is exercised, a Corbyn quart is sufficient. The date of collection and particulars of the rainfall should be noted at the time.

The results of analysis are most easily expressed in parts per 100,000, but as there are 70,000 grains to the gallon, the results may afterwards be expressed as grains per gallon by decreasing them in the proportion of 10 to 7—that is to say, by dividing by 10 and multiplying by 7. See Table IV., p. 39.

24. The order of work in the analysis of the water should be somewhat as follows :

First enter particulars of source of water, etc., in the analysis-book, noting the appearance of the sample, the amount of suspended matter, and any smell. The sample should be shaken up, and the actual analysis should then take place in the following order, so as to take up a minimum amount of time :

    i. *For the Total Solids.*—Start 100 c.c. to evaporate down to dryness on the water-bath.

    ii. *For the Free and Organic Ammonia.*—Start the distillation of 500 c.c. of the water.

    iii. *For the Estimation of Nitrates.*—Set 10 c.c. of the water and 10 c.c. of the standard nitrate solution to evaporate *nearly* to dryness by placing the two beakers on the top of the water-bath.

    iv. *Estimate the Chlorine* as chlorides in 100 c.c. of the water.

    v. *Finish the Estimation of the Free and Organic Ammonia.*

    vi. *Finish the Estimation of the Nitrates.*

    vii. *Finish the Estimation of the Total Solids.*

    viii. *Examine any Deposit Microscopically.*

ix. When it is desired to *estimate the Oxygen consumed* in four
hours, this process should be started even before the
total solids.

x. When the *Hardness* is to be estimated, it is best to do it
after the total solid matters have been determined, as
the results obtained will, as a rule, be somewhat of a
guide.

TOTAL SOLIDS.

25. Having thoroughly cleaned, dried, and made a note of the
tare weight of the nickel dish as described (sec. 3), put it on the
water-bath and light the gas.   Shake the sample of water up and
measure 100 c.c. in the measuring-flask ; pour this carefully into the
nickel dish, and leave it on the water-bath until dry ; then remove
it with a pair of crucible tongs, wipe the outside of the dish so that
no dirt from the water-bath adheres to it,* then transfer it to the
steam-oven and leave it half an hour.   It must then be set aside to
cool in the desiccator.   The tare weight of the dish should be put
on the right-hand pan of the balance, and the dish and its contents
must be weighed *as rapidly as possible.*

The source of the water will give some indication as to the probable
extra weights which will be required.   Thus, water from moorland
gathering grounds will contain less than 20 parts per 100,000—that
is, will contain less than ·02 gramme in the 100 c.c.—so that with such
a water it will be safe to place on the right pan of the balance the
tare weight of the dish and ·01 gramme, the rider being placed at
the end of the beam also indicating 10 milligrammes (·01 gramme),
so that there would be on the right-hand side of the balance
·02 gramme in addition to the tare weight of the dish.   The nickel
dish is then to be transferred from the desiccator to the left pan of
the balance, the screw turned, and the pans released.   If the weight
be too much, the rider should be placed at the 5 milligrammes, and
so on by halving until the correct weight is obtained.

If the water be derived from a well of which nothing is known,
before the dish is put upon the pan, ·05 should be placed on the
right side; if this proves too little, ·02 should be added ; if too
much, ·02 should be tried by itself and ·01 added or subtracted as
required, until the final adjustment can be made by the rider.
The glass front of the balance should be kept closed as much as
possible.

The gross weight should then be entered in the weighing-book,

* This is practically prevented by using a porcelain ring on the water-bath.

with the number of the sample, over the top of the tare weight of the dish, which is then to be subtracted thus :

1899. MARCH 25TH. SAMPLE NO. 250, DERBY TAP WATER.

		Grammes.
Residue from 100 c.c. of water taken and dish weigh	...	34·959
Tare weight of dish	...	34·931
		·028

That is to say, the Derby tap water contains :

·028 gramme in 100 c.c., or
·28 gramme in 1,000 c.c., or
2·8 grammes in 10,000 c.c., or
28·0 grammes in 100,000 c.c.

That is, 28 parts of solid matter in the 100,000.

It is not necessary to write every calculation out in this way. It will be sufficient to remember that when 100 c.c. of the water are taken the number of milligrammes gives the parts per 100,000. If the result be wanted in grains per gallon, multiply the result obtained by 7 and divide by 10, or see Table IV., p. 39.

The total amount of solid matter in a water depends upon its source. With surface waters it is, as a rule, 5 to 20 parts per 100,000, depending upon the geological formation, while well waters as a rule contain from 20 to 60, and even more, solids. The author has found that the average solids in some hundreds of sewages is 140 parts per 100,000. In every case a knowledge of the geological formation from which the water is derived is necessary, as explained under the heading of " Interpretation of the Results of Analysis."

If it be required to know the solids in suspension, first estimate the solids in 200 c.c. of the water, by adding 100 c.c. to the dish in which the first 100 c.c. have been evaporated down ; the weight given by the 200 c.c. should be double that of the 100. Then filter the water through filter-paper, and estimate the solids in 200 c.c. of the filtered water. The difference between the two weights will be due to the weight of the suspended solids in 200 c.c. of the water.

ESTIMATION OF THE FREE AND ORGANIC (ALBUMINOID) AMMONIA,

26. Having cleansed the distillation-flask with a little hydrochloric acid (sec. 7), it should be well rinsed out to get rid of all acid. Fill the 500 c.c. measuring-flask (sec. 11) up to the mark with the water to be analyzed, and pour this carefully through the larger funnel into the distillation-flask. Wash the thistle funnel and tin pipe by passing clean water through them, or by letting the tap water flow through. Then clean the tin pipe of the condenser in a similar way.

The distillation-flask is then to have its bung, which should be rinsed clean, inserted into it ; then its tin pipe is to be fitted into the block-tin pipe of the condenser, and the flask is to be clamped into position, the condenser being fixed at such a height that the lower end of it permits the Nessler glasses in their stand going under it. If the water is acid, it should be rendered alkaline by a little freshly-burnt carbonate of soda crystals dissolved in ammonia-free water. All now being ready, the water is turned on through the condenser, the Nessler tube marked 1 brought under the block-tin pipe of the condenser, and the lighted safety Bunsen burner brought direct under the flask. To prevent the flask breaking, two precautions must be taken : first, the outside is to be wiped dry before applying the naked flame; second, the flame must never be allowed to play upon any part of the flask above the level of the water, therefore at the end of the process, when there is only about 100 c.c. left in the flask, the flame must be lowered.

As soon as tube No. 1 is full up to the 100 c.c. mark, the stand should be moved along for the condenser to deliver into tube 2. As soon as this is full, 2 c.c. of Nessler's test (sec. 16, b) should be dropped into it and the stand moved on to the blank tube. If tube No. 2 gives no yellow tint, all the free ammonia has come off in the first tube, and the 50 c.c. alkaline permanganate (sec. 16, c) may be poured in. If tube 2 shows a trace of ammonia, the blank tube must be allowed to collect about 50 c.c. more ; that is, to be half filled, and to this 2 c.c. of Nessler's test are to be added. If the test shows that ammonia is still being given off, this must be poured into the 500 c.c. flask, and more collected, until all the ammonia has come off.

As soon as no colour is given with the Nessler test, 50 c.c. of the alkaline permanganate are to be measured in the clean measuring glass (sec. 11), and poured gently through the thistle funnel into the distillation flask, tube No. 3 having been previously placed under the condenser to collect the organic ammonia. As soon as tube 3 is full, tube 4 is to be brought under, and as soon as this is half full it is to be Nesslerized. If it contain no ammonia, as shown by there being no colour on the addition of the Nessler's test, the gas may be turned off, as the process is complete ; if it does contain ammonia, more distillate is to be collected in the spare tube, the contents of which, if they contain any free ammonia, are to be first transferred to the 500 c.c. flask for estimation afterwards. As a rule, all the free ammonia comes off in the first 200 c.c., and all the organic ammonia in the first 150 c.c., after the alkaline permanganate has been added ; as a rule, therefore, 150 c.c. are left in the flask—that is, 100 c.c. of the original water and 50 c.c. derived from the alkaline permanganate added. If more than 250 are distilled off for the free ammonia and more than 200 for the organic ammonia, the flame must be lowered so that it does not play upon the glass above the water level, and, if necessary, 50 or 100 c.c. of ammonia-free water should be poured through the thistle funnel so that the distillation may be continued further.

When the free and organic ammonia are all off, each tube has to be Nesslerized, and the depth of colour has to be matched with standard ammonia. This is done in the following manner : 2 c.c. of Nessler's test are dropped into each tube, which should be allowed to stand two minutes. With a little experience an approximate idea will be obtained as to the probable amount of standard ammonia solution required to match the depth of colour. Suppose it is decided to try 5 c.c., this quantity of the weak standard ammonia solution (sec. 16, *a*) is pipetted into the graduated Nessler glass ; this is then diluted with ammonia-free distilled water up to about 90 c.c., then 2 c.c. of Nessler are added, and it is made up to the 100 c.c. mark with ammonia-free distilled water ; this tube, which we shall call the " standard 5," is then left for two minutes.

If tube 1 is lighter, the standard solution should be poured into the graduated cylinder (sec 11) until the shades match, and the quantity of the standard left is to be read off. If 80 c.c. of the 100 c.c. are required to match the tint tube 1 is equal to $\frac{80}{100}$ of standard 5 c.c.—that is, standard 4 c.c. and tube No. 1 will contain ·00004 gramme of ammonia ; then tube 2 must be estimated in the same manner. If the tube 1 is darker than standard 5, it should be diluted with tube 2 and the two tubes then be estimated. If they are still too dark, as will be the case with sewage, they can be poured into the 500 c.c. flask and diluted with ammonia-free distilled water to, say, 400 c.c. The distillate and the ammonia-free water are thoroughly mixed, and the quantity is measured, which gives the same shade as the standard 5, from which the whole of the free ammonia in the 400 c.c. can be estimated. Thus, suppose 80 c.c. of the 400 c.c. equal the standard 5, a simple proportion sum shows the whole to equal a standard 25.

The first thing, therefore, is to ascertain what amount of standard solution the free ammonia is equal to : call this x ; then, knowing that 1 c.c. of standard solution equals ·00001 gramme of ammonia, the free ammonia in the 500 c.c. of water equals ·00001 multiplied by x grammes.　　　　　　　　　　　　　　　　　　　　•

Thus, if the first tube equals standard 5, and the second tube equals standard 2, the free ammonia in the 500 c.c. of water equals standard 7. But standard 1 contains ·00001 gramme, therefore standard 7 contains ·00007 gramme. Therefore 500 c.c. of water contains ·00007 gramme of ammonia.

			Gramme of Ammonia.

Then multiply by 2. This gives 1,000 c.c. of water containing ·00014
Then multiply by 10. „ 10,000 „ „ ·0014
Then multiply by 10 again. „ 100,000 „ „ ·014
That is, the free ammonia is ·014 per 100,000.

The organic ammonia is estimated in precisely the same manner. For the analysis of sewage, 25 c.c. should be accurately measured with a pipette and transferred to the distillation-flask, 500 c.c. of ammonia-free distilled water being added.

In this case, as a rule, the free ammonia need not be estimated, as it is of no practical importance, some of the best effluents containing as much as the worst. It is best to collect the free ammonia of a sewage in a 500 c.c. flask until it is proved (by taking a small quantity in a Nessler tube) that it has all come off. If it is desired to estimate the free ammonia, as a rule it will be necessary to dilute the distillate with ammonia-free distilled water to 500 c.c. in order to get a sufficient dilution for a suitable quantity to be equal to 5 c.c. of the standard ammonia.

The organic ammonia in a sewage or sewage effluent can be collected in 100 c.c. Nessler tubes, and be diluted with distilled water afterwards if the colour proves too dark for estimation. The very greatest care must be exercised to see, when dilutions are made, that the distilled water added, and the distillate, are thoroughly mixed.

In this case the calculation is as follows :

Suppose the free ammonia is equal to a standard 45, and the organic ammonia equal to a standard 7 : 25 c.c. of sewage contains as much free ammonia as standard 45.

Standard 1 equals ·00001 gramme of ammonia; standard 45 equals ·00045 gramme of ammonia.
Therefore 25 c.c. of sewage contains ·00045 gramme of ammonia.
Multiplying by 4: 100 c.c. of sewage contains ·0018 gramme of ammonia.
„ 10: 1,000 „ „ ·018 „ „
„ 10: 10,000 „ „ ·18 „ „
„ 10: 100,000 „ „ 1·8 grammes „

The sewage therefore contains 1·8 parts per 100,000 of free ammonia.

And for the organic ammonia, 25 c.c. of sewage contains as much ammonia as standard 7.

Standard 1 equals ·00001 gramme of ammonia; standard 7 equals ·00007 gramme of ammonia.
Therefore 25 c.c. of sewage contains ·00007 gramme of ammonia.
Multiplying by 4: 100 c.c. of sewage contains ·00028 gramme of ammonia.
„ 10: 1,000 „ „ ·0028 „ „
„ 10: 10,000 „ „ ·028 „ „
„ 10: 100,000 „ „ ·28 „ „

The sewage therefore contains 0·28 part of organic ammonia in the 100,000.

To save time the following table can be used, giving the ammonia in parts per 100,000 for each c.c. of standard solution used, when 500 c.c., 100 c.c., or 25 c.c. of the water or sewage are taken for analysis :

TABLE I.—Giving the Ammonia in Parts per 100,000 for each c.c. of Standard Ammonia used.

Cubic Centimetres of Standard Ammonia Solution (1 c.c. = ·00001 grm. of NH₃).	NH₃ IN PARTS PER 100,000.			Cubic Centimetres of Standard Ammonia Solution (1 c.c. = ·00001 grm. of NH₃).	NH₃ IN PARTS PER 100,000.		
	Waters : 500 c.c. taken.	Sewage Effluents : 100 c.c. taken.	Sewages : 25 c.c. taken.		Waters : 500 c.c. taken.	Sewage Effluents : 100 c.c. taken.	Sewages : 25 c.c. taken.
1·0	0·002	0·01	0·04	11·0	0·022	0·11	0·44
1·25	0·0025	0·0125	0·05	11·5	0·023	0·115	0·46
1·5	0·008	0·015	0·06	12·0	0·024	0·12	0·48
1·75	0·0035	0·0175	0·07	12·5	0·025	0·125	0·50
2·0	0·004	0·02	0·08	13·0	0·026	0·13	0·52
2·25	0·0045	0·0225	0·09	13·5	0·027	0·135	0·54
2·5	0·005	0·025	0·10	14·0	0·028	0·14	0·56
2·75	0·0055	0·0275	0·11	14·5	0·029	0·145	0·58
3·0	0·006	0·03	0·12	15·0	0·030	0·15	0·60
3·25	0·0065	0·0325	0·13	15·5	0·031	0·155	0·62
3·5	0·007	0·035	0·14	16·0	0·032	0·16	0·64
3·75	0·0075	0·0375	0·15	16·5	0·033	0·165	0·66
4·0	0·008	0·04	0·16	17·0	0·034	0·17	0·68
4·25	0·0085	0·0425	0·17	17·5	0·035	0·175	0·70
4·5	0·009	0·045	0·18	18·0	0·036	0·18	0·72
4·75	0·0095	0·0475	0·19	18·5	0·037	0·185	0·74
5·0	0·010	0·05	0·20	19·0	0·038	0·19	0·76
5·25	0·0105	0·0525	0·21	19·5	0·039	0·195	0·78
5·5	0·011	0·055	0·22	20·0	0·040	0·20	0·80
5·75	0·0115	0·0575	0·23	21·0	0·042	0·21	0 84
6·0	0·012	0·06	0·24	22·0	0·044	0·22	0·88
6·25	0·0125	0·0625	0·25	23·0	0·046	0·23	0·92
6·5	0·013	0·065	0·26	24·0	0·048	0·24	0·96
6·75	0·0135	0·0675	0·27	25·0	0 050	0·25	1·00
7·0	0·014	0·07	0·28	26·0	0·052	0·26	1·04
7·25	0·0145	0·0725	0·29	27·0	0·054	0·27	1 08
7·5	0·015	0·075	0·30	28·0	0·056	0·28	1·12
7·75	0·0155	0·0775	0·31	29·0	0·058	0·29	1·16
8·0	0·016	0·08	0·32	30·0	0·060	0·30	1·20
8·25	0·0165	0·0825	0·33	31 0	0·062	0·31	1·24
8·5	0·017	0·085	0·34	32·0	0·064	0·32	1·28
8·75	0·0175	0·0875	0·35	33·0	0·066	0·33	1·32
9·0	0·018	0·09	0·36	34·0	0·068	0·34	1·36
9·25	0·0185	0·0925	0·37	35·0	0·070	0·35	1·40
9·5	0·019	0·095	0·38	40·0	0·080	0·40	1·60
9·75	0·0195	0·0975	0·39	45·0	0·090	0·45	1·80
10·0	0·020	0·10	0·40	50·0	0·100	0·50	2·00
10·5	0·021	0·105	0·42				

The author has worked with as small a quantity as 25 c.c. of sewage now for some years, and has found that perfectly accurate results can be obtained if the water added is absolutely free from ammonia, and thorough mixing of the distilled water and the distillate is made a point of, before Nesslerizing.

ESTIMATION OF THE CHLORINE.

27. The chlorides are estimated as chlorine in the following manner : The 100 c.c. measuring-flask is filled up to the mark with the water to be examined ; this is then decanted into the 200 c.c. porcelain dish, and two spots of the 10 per cent. monochromate of potash solution (sec. 17, *b*) are added so as to give the water in the dish a canary-yellow colour. The nitrate of silver burette is then to be filled to the 0 mark with the standard nitrate of silver solution (sec. 17, *a*), 1 c.c. of which precipitates ·001 gramme of chlorine. The standard nitrate of silver solution is then carefully run into the dish, which is stirred with a clean glass rod until all the chlorine is precipitated. This is known by the nitrate of silver then forming the red chromate of silver with the chromate of potash. As soon as there is the slightest red tinge, which does not disappear on stirring, the amount of standard nitrate of silver used should be read off. The result is the number of milligrammes (·001 gramme) in 100 c.c., and therefore the number of parts per 100,000. The reaction is a very delicate one, and at the end the nitrate of silver should be added $\frac{1}{10}$ c.c. at a time, the estimation being made twice over if necessary.

The amount of chlorine in drinking waters varies with the nature of the geological formation from which they are derived, and a knowledge of the isochlor or the normal chlorine of the unpolluted water is necessary before any conclusions can be drawn. This subject is dealt with in the chapter on " The Interpretation of the Results of Analysis." It may be said here, however, that good surface waters contain less than 2 parts per 100,000 as a rule, while well waters may contain any proportion, depending upon the formation. If the normal chlorine be known, no well water should be passed with a larger amount of chlorine until a local investigation proves the chlorine to be of mineral origin.

NITROGEN AS NITRATES AND NITRITES.

28. (*a*) Measure 10 c.c. of the water to be examined in a 10 c.c. pipette (sec. 13), and run it out into a small beaker (sec. 14). Wash the pipette through with distilled water, and then measure 10 c.c. of the nitrate of potash solution (sec. 18, *a*) into another small beaker, carefully making a written note of the number of the beaker in which each is contained. Place the two beakers on the upper part of the water-bath at the same time as the total solids are being evaporated down. Take them off as soon as they are dry, or there may be loss of nitrate of ammonia. When they are both cold, add to each 1 c.c. (measured in the 1 c.c. pipette, sec. 13) of phenol sulphuric acid (sec. 18, *b*). Put them again on the water-bath for ten minutes. Then wash each beaker out carefully into a Nessler tube, using successive quantities of water, until each tube is half full; then add to each 25 c.c. of liquor potassæ (measured in the graduated 50 c.c. cylinder, sec. 11), and fill up each tube to the 100 c.c. mark.

Any nitrates present convert the phenol sulphuric acid into picric acid, which by the action of the caustic potash forms picrate of potash, the intensity of the colour depending upon the amount of nitrates present.

The standard is equal to 1 part per 100,000. If the Nessler glass containing the water being examined is darker, measure the quantity equal to the standard tube; if 80 is equal to it, then the quantity present is as follows: $80 : 100 :: 1 : x = 1\cdot25$; the nitrogen as nitrates being $1\cdot25$ parts per 100,000.

If the tube is lighter than the standard, pour off the standard until the tint is matched; then if 75 c.c. are left in the standard tube, the nitrogen as nitrates is $\cdot75$ part per 100,000.

The following table will be found useful for rapidly ascertaining the quantity of nitrates present :

TABLE II.—Giving Parts per 100,000 of Nitrogen as Nitrates for each c.c. of Standard Solution.

No. of c.c. of Yellow Solution equal to the Standard 100 c.c.	Nitrogen as Nitrates : Parts per 100,000.	No. of c.c. of Yellow Solution equal to the Standard 100 c.c.	Nitrogen as Nitrates : Parts per 100,000.
100	1·00	50	2·00
95	1·05	48	2·08
90	1·11	46	2·17
85	1·18	45	2·22
80	1·25	44	2·28
78	1·28	42	2·38
76	1·32	40	2·50
75	1·33	38	2·63
74	1·35	36	2·78
72	1·39	35	2·86
70	1·43	34	2·94
68	1·47	32	3·13
66	1·51	30	3·33
65	1·54	28	3·57
64	1·55	26	3·85
62	1·61	25	4·00
60	1·67	24	4·17
58	1·72	22	4·55
56	1·78	20	5·00
55	1·82	18	5·55
54	1·85	16	6·25
52	1·92	15	6·67

(*b*) It is always as well to make a qualitative test for the presence of nitrites by adding 5 spots of the iodide of potash solution (sec. 21, *d*) and 5 spots of the starch solution (sec. 21,.*c*) to about 25 c.c. of the water in a Nessler tube or in a test-tube, mixing well, and adding about 10 c.c. of the dilute sulphuric acid (sec. 21, *c*). If nitrites are present the sulphuric acid disengages the nitrous acid, which liberates iodine from the iodide of potash, and this in turn strikes a blue colour with the starch. The blue colour should come *immediately* if an appreciable amount of nitrite is present. The test should be done in duplicate, one tube containing as a control a good tap-water free from nitrites.

HARDNESS.

29. (*a*) Having previously determined the total solids, the amount of water to take for the estimation of hardness will be known.

If the total solids are less than 20 per 100,000, take 50 c.c.

If the total solids are between 20 and 50 per 100,000, take 25 c.c.

If the total solids are above 50 per 100,000, take 10 c.c.

In the two latter cases make up to 50 c.c. with recently-boiled and cooled distilled water.

Put the water in a 6-ounce stoppered bottle, thoroughly shake, and then draw out the disengaged carbonic acid through a glass tube. Add carefully the soap solution (sec. 20) from a burette, 1 c.c. at a time, and shake vigorously after each addition until a thin uniform lather is formed, which will last for five minutes when the bottle is placed upon its side. If the lather breaks add a drop more soap solution, the end of the reaction being determined by adding one-tenth of a cubic centimetre at a time.

Read off the cubic centimetres of soap solution used, and refer to the following table, which is also printed on the bottle containing the soap solution, to ascertain the hardness equivalent to the cubic centimetres of soap solution used. When 50 c.c. of the water are used, the figure obtained is the total hardness in parts per 100,000. When 25 c.c. are used, the hardness obtained from the table must be multiplied by 2, and when 10 c.c. are used it must be multiplied by 5. This is the total hardness.

(*b*) To ascertain the permanent hardness, boil a known quantity of the water, say 100 c.c., briskly in the porcelain dish for fifteen minutes; then filter into the 100 c.c. flask, make up with the distilled water for the water evaporated, and determine the hardness in 50 c.c. This is the permanent hardness, and the temporary or removable hardness is the difference between this and the total hardness.

TABLE III.

Hardness of Water (50 c.c. used).

Volume of Soap Solution.	CaCO₃ per 100,000.	Degrees of Hardness.	Volume of Soap Solution.	CaCO₃ per 100,000.	Degrees of Hardness.	Volume of Soap Solution.	CaCO₃ per 100,000.	Degrees of Hardness.
c.c.			c.c.			c.c.		
0·7	0·00	0·00	5·9	7·29	5·10	11·1	15·00	10·50
0·8	0·16	0·11	6·0	7·43	5·20	·2	15·16	10·61
0·9	0·32	0·22	·1	7·57	5·80	·3	15·32	10·72
1·0	0·48	0·34	·2	7·71	5·40	·4	15·48	10·84
·1	0·63	0·44	·3	7·86	5·50	·5	15·63	10·94
·2	0·79	0·55	·4	8·00	5·60	·6	15·79	11·05
·3	0·95	0·67	·5	8·14	5·70	·7	15·95	11·17
·4	1·11	0·78	·6	8·29	5·80	·8	16·11	11·28
·5	1·27	0·89	·7	8·43	5·90	·9	16·27	11·39
·6	1·43	1·00	·8	8·57	6·00	12·0	16·43	11·50
·7	1·56	1·09	·9	8·71	6·10	·1	16·59	11·61
·8	1·69	1·18	7·0	8·86	6·20	·2	16·75	11·73
·9	1·82	1·27	·1	9·00	6·30	·3	16·90	11·83
2·0	1·95	1·37	·2	9·14	6·40	·4	17·06	11·94
·1	2·08	1·46	·3	9·29	6·50	·5	17·22	12·05
·2	2·21	1·55	·4	9·43	6·60	·6	17·38	12·17
·3	2·34	1·64	·5	9·57	6·70	·7	17·54	12·28
·4	2·47	1·73	·6	9·71	6·80	·8	17·70	12·39
·5	2·60	1·82	·7	9·86	6·90	·9	17·86	12·50
·6	2·73	1·91	·8	10·00	7·00	13·0	18·02	12·61
·7	2·86	2·00	·9	10·15	7·11	·1	18·17	12·72
·8	2·99	2·09	8·0	10·30	7·21	·2	18·33	12·83
·9	3·12	2·18	·1	10·45	7·32	·3	18·49	12·94
3·0	3·25	2·28	·2	10·60	7·42	·4	18·65	13·06
·1	3·38	2·37	·3	10·75	7·53	·5	18·81	13·17
·2	3·51	2·46	·4	10·90	7·63	·6	18·97	13·28
·3	3·64	2·55	·5	11·05	7·74	·7	19·13	13·39
·4	3·77	2·64	·6	11·20	7·84	·8	19·29	13·50
·5	3·90	2·73	·7	11·35	7·95	·9	19·44	13·61
·6	4·03	2·82	·8	11·50	8·05	14·0	19·60	13·72
·7	4·16	2·91	·9	11·65	8·16	·1	19·76	13·83
·8	4·29	3·00	9·0	11·80	8·26	·2	19·92	13·94
·9	4·43	3·10	·1	11·95	8·37	·3	20·08	14·06
4·0	4·57	3·20	·2	12·11	8·48	·4	20·24	14·17
·1	4·71	3·30	·3	12·26	8·58	·5	20·40	14·28
·2	4·86	3·40	·4	12·41	8·69	·6	20·56	14·39
·3	5·00	3·50	·5	12·56	8·79	·7	20·71	14·50
·4	5·14	3·60	·6	12·71	8·90	·8	20·87	14·61
·5	5·29	3·70	·7	12·86	9·00	·9	21·03	14·72
·6	5·43	3·80	·8	13·01	9·11	15·0	21·19	14·83
·7	5·57	3·90	·9	13·16	9·21	·1	21·35	14·95
·8	5·71	4·00	10·0	13·31	9·32	·2	21·51	15·06
·9	5·86	4·10	·1	13·46	9·42	·3	21·68	15·18
5·0	6·00	4·20	·2	13·61	9·53	·4	21·85	15·30
·1	6·14	4·30	·3	13·76	9·63	·5	22·02	15·41
·2	6·29	4·40	·4	13·91	9·74	·6	22·18	15·53
·3	6·43	4·50	·5	14·06	9·84	·7	22·35	15·65
·4	6·57	4·60	·6	14·21	9·95	·8	22·52	15·76
·5	6·71	4·70	·7	14·37	10·06	·9	22·69	15·88
·6	6·86	4·80	·8	14·52	10·16	16·0	22·86	16·00
·7	7·00	4·90	·9	14·68	10·28			
·8	7·14	5·00	11·0	14·84	10·39			

DETERMINATION OF THE OXYGEN CONSUMED.

30. In this process it is necessary to make a control determination with recently-boiled distilled water, as the hyposulphite solution is extremely unstable. The time during which the process is continued, namely, one, three, or four hours, and the temperature, 80° F., or boiling-point, is somewhat a matter of fashion. Thresh recommends boiling for fifteen minutes; Tidy recommended one hour and three hours at 80° F.; at the present time, four hours at 80° F. is the most usual time. It is obvious that, unless the time and temperature are fixed, the results of this process can never be generally adopted.

Cleanse two 12-ounce stoppered bottles by rinsing round with strong sulphuric acid. Into 1 place 250 c.c. of pure distilled water, and in the other, which we will call 2, pour 250 c.c. of the water to be examined; into each pipette 10 c.c. of the dilute sulphuric acid (sec. 21, c), and then 10 c.c. of the permanganate solution (sec. 21, a); allow them to stand at the temperature decided upon, and as often as the permanganate solution in the bottle 2 fades add another 10 c.c. of standard permanganate (sec. 21, a).

At the expiration of the time fixed upon add 2 c.c. of the iodide solution (sec. 21, d). The pink colour of the permanganate is thereby changed to yellow by the oxygen of the permanganate liberating an equivalent amount of iodine.

Now add from a burette the thiosulphate solution (sec. 21, b), until the yellow turns a faint straw colour; then add 1 c.c. of starch solution (sec. 21, e): a deep blue colour appears; add the thiosulphate gradually until the blue colour absolutely disappears. Make a note of the thiosulphate used, and proceed in the same way with the other bottle.

With the distilled water the 10 c.c. of the permanganate should remain unchanged, so that if x c.c. of thiosulphate are used, these x c.c. correspond to 10 c.c. of the permanganate solution—that is, 1 milligramme of oxygen (sec. 21). If the water, being analyzed, takes only y c.c. of the thiosulphate, the oxygen consumed by the oxidizable matter in the water is $1 - \dfrac{y}{x}$ milligrammes. If a second 10 c.c. of permanganate had been required by the water in question, the oxygen consumed is $2 - \dfrac{y}{x}$ milligrammes. To express the amount of oxygen consumed in parts per 100,000, it will be necessary to multiply the result by 4, as 250 c.c.—that is, a quarter of a thousand—were used. In the case of sewages, it will be sufficient to take 50 c.c., and the results in this case must be multiplied by 20 instead of 4 to be expressed in parts per 100,000.

Estimation of Lead.

31. As a rule, the waters which act most dangerously upon lead are moorland waters off peaty gathering grounds, chiefly when a heavy rain comes on after a drought, and there is much decaying peat and vegetable matter upon the collecting area.

A water which is sufficiently acid to give a red colour with methyl orange (sec. 23) will almost invariably act upon lead pipes. Such a water may be acid as it enters the storage reservoirs and distribution-pipes, but will no longer be acid as it comes from the taps, it having neutralized its acidity by attacking the iron and lead of the pipes.

To estimate the amount of lead in a water, take two Nessler tubes, put 100 c.c. of distilled water into the one, and 100 c.c. of the water to be examined into the other. If the latter is brown from peat, add a little burnt sugar or some vegetable colour to the distilled water to match the tint. Then add a few drops of acetic acid to the water to be examined, and then 1 c.c. of ammonium sulphide (sec. 22, *b*), measured in the 1 c.c. pipette, and stir with a glass rod. If lead is present a brown colour will appear. To estimate the quantity of lead, add 1 c.c. of ammonium sulphide to the distilled water, and drop into it by means of the graduated 5 c.c. pipette (Fig. 14) the standard lead solution, a spot at a time, carefully stirring all the time. As soon as the depth of colour is matched note the quantity of lead solution used, 1 c.c. being equivalent to ·0001 gramme. If this is the quantity required to equal the depth of colour in the 100 c.c. of the water, the lead present is ·1 part per 100,000; if 1¼ c.c. are used, it will be 1·25 parts per 100,000, and so on, using either greater or smaller quantities until the colour is exactly matched. ·05 part of lead per 100,000 and over is a positively dangerous amount.

To Express Results in Grains per Gallon.

32. If it is desired to express the results in grains per gallon, the following table will be found useful in preventing unnecessary calculations :

TABLE IV.

For the Conversion of Parts per 100,000 into Grains per Gallon.

Parts per 100,000.	Grains per Gallon.	Parts per 100,000.	Grains per Gallon.	Parts per 100,000.	Grains per Gallon.	Parts per 100,000.	Grains per Gallon.
1	0·7	48	33·6	95	66·5	142	99·4
2	1·4	49	34·3	96	67·2	143	100·1
3	2·1	50	35·0	97	67·9	144	100·8
4	2·8	51	35·7	98	68·6	145	101·5
5	3·5	52	36·4	99	69·3	146	102·2
6	4·2	53	37·1	100	70·0	147	102·9
7	4·9	54	37·8	101	70·7	148	103·6
8	5·6	55	38·5	102	71·4	149	104·3
9	6·3	56	39·2	103	72·1	150	105·0
10	7·0	57	39·9	104	72·8	151	105·7
11	7·7	58	40·6	105	73·5	152	106·4
12	8·4	59	41·3	106	74·2	153	107·1
13	9·1	60	42·0	107	74·9	154	107·8
14	9·8	61	42·7	108	75·6	155	108·5
15	10·5	62	43·4	109	76·3	156	109·2
16	11·2	63	44·1	110	77·0	157	109·9
17	11·9	64	44·8	111	77·7	158	110·6
18	12·6	65	45·5	112	78·4	159	111·3
19	13·3	66	46·2	113	79·1	160	112·0
20	14·0	67	46·9	114	79·8	161	112·7
21	14·7	68	47·6	115	80·5	162	113·4
22	15·4	69	48·3	116	81·2	163	114·1
23	16·1	70	49·0	117	81·9	164	114·8
24	16·8	71	49·7	118	82·6	165	115·5
25	17·5	72	50·4	119	83·3	166	116·2
26	18·2	73	51·1	120	84·0	167	116·9
27	18·9	74	51·8	121	84·7	168	117·6
28	19·6	75	52·5	122	85·4	169	118·3
29	20·3	76	53·2	123	86·1	170	119·0
30	21·0	77	53·9	124	86·8	171	119·7
31	21·7	78	54·6	125	87·5	172	120·4
32	22·4	79	55·3	126	88·2	173	121·1
33	23·1	80	56·0	127	88·9	174	121·8
34	23·8	81	56·7	128	89·6	175	122·5
35	24·5	82	57·4	129	90·3	176	123·2
36	25·2	83	58·1	130	91·0	177	123·9
37	25·9	84	58·8	131	91·7	178	124·6
38	26·6	85	59·5	132	92·4	179	125·3
39	27·3	86	60·2	133	93·1	180	126·0
40	28·0	87	60·9	134	93·8	181	126·7
41	28·7	88	61·6	135	94·5	182	127·4
42	29·4	89	62·3	136	95·2	183	128·1
43	30·1	90	63·0	137	95·9	184	128·8
44	30·8	91	63·7	138	96·6	185	129·5
45	31·5	92	64·4	139	97·3	186	130·2
46	32·2	93	65·1	140	98·0	187	130·9
47	32·9	94	65·8	141	98·7	188	131·6

CHAPTER V.

The Interpretation of the Results of Analysis.

33. Before the results of the analysis of a water can be correctly interpreted, it must first be known whether the sample is one of river water, surface water, or water from a spring or deep well. A correct description of its source and surroundings should be obtained if possible by a careful personal inspection, and before deciding whether the results of any given analysis indicate pollution they should be compared with the results obtained by the analysis of an obviously unpolluted water of the same character and from a similar source.

The values to be attached to the various results obtained in the case of river waters differ from those obtained by the analysis of well or spring waters, and although certain general conclusions may safely be drawn in each class of water, yet as a rule it is necessary to have normal local standards with which to compare our figures. Every village will have its own local standards, and even several sets of these, where the geological formations are much broken up. Where the local normal chlorine, the normal nitrates, and the normal solid matter are known, any serious departure from them will indicate pollution.

We will now briefly consider the conclusions to be drawn from the amount of the solid matter, the chlorine, and the other figures obtained upon analysis.

34. *Total Solids.*—The solid matter in different waters varies generally with their sources. Upland surface waters contain but small quantities, the actual amount depending upon the solubility of the rocks forming the gathering ground. Springs and deep wells contain much more, as the water has been in contact with the rocks to a much greater extent.

The following are the results which may be expected from the various geological formations, as given by the Rivers Pollution Commissioners :

TABLE V.

Unpolluted Upland Surface Waters.

TOTAL SOLIDS.

Geological Formation.	Average.	Highest.	Lowest.
Igneous and metamorphic rocks	5	12	2
Yoredale and millstone grits, and the non-calcareous portions of the coal-measures	9	15	5
The Lower London tertiaries and Bag-shot beds	8·4	14	6
The calcareous portions of the silurian ...	14	15	12
The mountain limestone	17	24	14
The calcareous portions of the coal-measures	23	50	12
The lias, new red sandstone, and magnesian limestone	19	25	11

If possible, a sample of unpolluted water from the identical gathering ground should be obtained as a standard of normal solids.

Unpolluted Spring Waters.

TOTAL SOLIDS.

Geological Formation.	Average.	Highest.	Lowest.
Granite and gneiss	6	10	3·5
Silurian rocks	12·5	26	8
Devonian rocks and old red sandstone ...	25	72	5
Mountain limestone	82	98	15
Yoredale and millstone grits	22	40	4
Magnesian limestone ... ... · ...	45	66*	85
New red sandstone	29	60	13
Lias	86	60	21
Oolites	80	52	22
Greensands and weald clay	80	68	5
Chalk	80	40	—
Fluvio-marine drift and gravel	61	22	25

Deep Well Waters.

TOTAL SOLIDS.

Geological Formation.	Average.	Highest.	Lowest.
Devonian rocks and millstone grits ...	82	55	10
Coal-measures	83	144	88
Magnesian limestone	61	84	44
New red sandstone	30	60	15
Lias	70	84	57
Oolites	83	41	26
Chalk	86	50	23
Chalk below London clay	78	106	83
Drift ... ·	53	61	45

* This is the only sample analyzed by Commissioners.

3—3

It will be noticed that with all formations the total solids are least in the surface waters, and most in the well waters. In springs it is presumable that the soluble solid matters have gradually been washed out, so that the total solids are therefore less than in well waters.

Some of the older authorities were in the habit of classing waters containing more than 70 parts of solid matter per 100,000 as waters which should be condemned. This was absolutely unreasonable when we remember that each day a human being requires about 1 ounce of mineral salts, such as the chlorides, phosphates, and sulphates of sodium, potassium, and calcium, and that he will probably not drink more than a quart of water. It must be admitted that as long as the saline constituents of the water are natural to the strata, and do not interfere with the palatability of the water, they may even amount to 200 parts per 100,000 without being detrimental to health. In this case, however, it is important to know exactly what the saline matter is. For instance, sulphate of magnesium in such quantity would have its therapeutic effects. From the figures set out above, the importance of taking the source of the water into consideration will be apparent.

35. *Chlorides.*—The chlorine in unpolluted upland surface waters should be less than 2 parts per 100,000. In deep wells and springs in the igneous, silurian, Devonian rocks, the mountain limestone, the Yoredale and millstone grits, the magnesian limestone, the new red sandstone, the oolites and the chalk, it is as a rule less than 4 parts per 100,000. In the coal-measures it may be anything from 2 to 50 parts without the water being polluted; so also deep wells near the sea may show high chlorines by insuction of sea-water. The chalk beneath the London clay is highly saline, and occasionally the magnesian limestone is saline. The keuper marls, the greensands and the drifts may contain as much as 10, or even occasionally 20, parts of chlorine in the 100,000 without being organically polluted, and as extreme instances of the same conditions the saline springs of Droitwich, Stafford, and Cheshire may be instanced.

To correctly interpret the value of a chlorine determination, a series of isochlors or normal chlorines for each local formation must be established. Then any departure from the normal points to probable sewage pollution. Sewage contains 7 to 15 parts per 100,000 more chlorine than the drinking water of the neighbourhood. It is a particularly valuable determination in the case of surface waters, in which the normal chlorine is much more constant than in well waters.

A very elementary knowledge of geology is sufficient to expose the unwarrantable assumption of the Rivers Pollution Commissioners when they state that "it would only rarely lead to the rejection of good water if all samples containing such an excessive proportion of chlorine as 5 parts in the 100,000 were condemned as unfit for domestic use." It is against such unscientific rule-of-thumb dogmatism, which would condemn the unpolluted public supplies of important communities, that these notes are written.

36. *Free Ammonia.*—The free ammonia in upland surface waters should not be more than ·002 part per 100,000; in times of heavy flood it may temporarily increase to ·004, but this should only be a temporary increase. In the purest upland surface waters it is nil or ·001. In spring waters it should not be more than ·003, and in well waters less than ·005. A large quantity is present in the case of waters from new wells and in waters containing a trace of iron, such as waters from the coal-measures, the green sandstone, the chalk beneath the London clay, also from very deep wells and boreholes, in which instances it owes its presence to the reduction of nitrates by the iron lining of the borehole or the metalliferous strata through which the water percolates. When the free ammonia is more than ·005 per 100,000, the amount of organic ammonia becomes a matter of the greatest importance. As a rule, when the free ammonia owes its presence to the reduction of nitrates, it will be present in such a large quantity as ·02 or even ·1 part per 100,000, and by keeping the water, if any trace of nitrates is present the free ammonia will frequently be found to increase, while the nitrogen as nitrates will diminish.

37. *Organic Ammonia.*—In contrast with the total solids, with the chlorine, and with the free ammonia, more organic ammonia is permissible in an upland surface water than in spring waters and deep well waters. In spring waters the quantity should not exceed ·003 or ·004 part per 100,000, and in well waters it should not exceed ·005 or ·006; if it does, and in other respects the water is normal, it may only mean that vegetable matter has fallen into the well, such as decaying timber, or that algæ or other low forms of vegetable life are growing in the water. But if the chlorine is excessive or the water contains ·005 or more free ammonia, or the nitrogen as nitrates is more than ·5, the indication is that the oxidizable organic matter is not of a purely vegetable nature, and that some polluting matter has found its way into the water.

On the other hand, with upland surface waters, particularly in times of heavy rain, water from an unpolluted gathering ground may contain from ·008 to ·01 part per 100,000. When it contains as much as ·01, however, it is, as a rule, an indication that the water requires filtering. ·

Unfiltered upland surface waters yielding ·008 of organic ammonia must only be passed as fit for use without filtration when they contain little free ammonia, normal chloride for the geological formation, and an unappreciable amount of nitrogen as nitrates, say less than ·01 part per 100,000.

38. *The Nitrogen as Nitrates.*—In upland surface waters the nitrogen as nitrates should hardly be an appreciable quantity. As soon as the drainage from cultivated land finds its way into the water, nitrates begin to appear, so that if present beyond a mere trace in such quantities as ·03 or ·04, it is a sign that the water should be safeguarded by proper sand filtration.

In well waters the nitrogen as nitrates may be as much as ·4 or ·5 part per 100,000, and in deep well waters, particularly in the chalk, the old red sandstone, and the magnesian limestone, it may be even more, without the water being in any way dangerous.

The presence of an excessive amount of nitrates is generally held to indicate that any organic ammonia present is of animal origin.

As was pointed out, a qualitative test for nitrites should also always be made, as the presence of an appreciable amount of nitrites in any water other than a deep well or a new well water (in which they may be due to the reduction of nitrates) must be regarded as a sign of recent sewage contamination.

39. *The Oxygen consumed.*—Frankland and Tidy's standards for oxygen consumed are :

	Upland Surface Waters.	Other Waters.
High organic purity	0·1	0·05
Medium	0·1 to 0·3	0·05 to 0·15
Doubtful	0·3 to 0·4	0·15 to 0·2
Impure	0·4	over 0·2

But these hard and fast lines cannot be accepted. A moderate amount of organic ammonia may be shown to be of vegetable origin by a high oxygen absorbed and low nitrates and chlorine.

40. *The Degree of Hardness.*—As far as health is concerned, no ill consequences are likely to follow from any degree of hardness up to 40 or 50. Of course hard waters are objectionable for a public supply, particularly in manufacturing urban districts ; but for purely domestic purposes exception ought not to be taken to an unpolluted water merely on the grounds of hardness unless it is very excessive, or unless a much better water is at hand.

41. Besides making a careful chemical examination of the water, a microscopical examination should be made of any deposit which settles, as most valuable hints may thus be obtained from the presence of shreds of cotton, hair, epithelial cells, etc.

THE REPORT.

42. In framing a report upon the analysis of a water, brevity should be aimed at. If upon analysis there is no evidence of pollution, the report should certainly not state that " the water is one of great purity," for, as Sir Richard Thorne has recently written : " It may be doubted whether we are yet in possession of laboratory methods for the prompt and certain recognition of uniform purity and safety of water " (Annual Report of the Medical Officer of the Local Government Board, 1898). The form the report should take is rather that " the water shows no evidence of pollution," or that " as far as chemical analysis can show the water is free from evidence of contamination," leaving it open for the obvious inference that chemical analysis has definite limitations, and other investigations, such as an inspection of the source, should be made. When, however, comparison with an unpolluted source of water shows that a sample submitted for analysis is polluted, there is no reason why the language of the report should be guarded. Subject to the remarks in the preceding paragraphs, the following table is suggestive of what unpolluted water should be like :

PARTS PER 100,000.

	Total Solids.	Chlorine.	Free Ammonia.	Organic Ammonia.	Nitrogen as Nitrates.
Upland surface waters ...	2 to 20	1 to 2	·000 to ·003	·001 to ·006	Nil.

If the nitrogen as nitrates amounts to ·04 it is a distinct indication that the water should be filtered. The solid matter and the chlorine will vary with the geological formation.

In peaty waters the organic ammonia may be ·01 or more, but in this case there should be no nitrogen as nitrates, the chlorine should be low, and the water probably of a brown colour, and perhaps of an acid reaction.

With spring waters the various determinations should be approximately as follows :

PARTS PER 100,000.

	Total Solids.	Chlorine.	Free Ammonia.	Organic Ammonia.	Nitrogen as Nitrates.*
Spring waters ...	10 to 50	2 to 4	less than ·002	·001 to ·004	·3

Here again all depends upon the geological formation, and the normal solids, chlorine, nitrates must be known before any inference can be drawn.

With regard to well waters, the solids, the chlorine, and the nitrogen as nitrates, will probably all be higher than in spring water ; the organic ammonia should be less than ·008 per 100,000. If the free ammonia, chlorine and nitrates are small, and the

* The nitrates may be much higher in springs in the chalk.

organic ammonia is ·008 to ·01, the indication is that some wood is rotting in the water ; this opinion would be confirmed by the oxygen absorbed being high, and the report should suggest that the well should be examined.   A high free ammonia and a high chlorine and a moderate amount of organic ammonia frequently is caused by pollution with urine.

In every case an unpolluted sample of water from the same geological formation should be obtained, with which standard the water should be compared.

As illustrating the method of working out isochlors, etc., I may quote the following analyses of waters from a small urban district with a population of 8,500.   The district is built upon a hill of Bunter sandstone dipping towards a brook.   To obtain normal standards a well at the top of the hill (No. 1), another well also at the top of the hill, but further removed, and a spring flowing into the brook a quarter of a mile before it reaches the township, were taken.   From these three samples it is plain the normal solids are from 85 to 40, the normal chlorines between 1 and 2, and the nitrates ·1 per 100,000.

The next samples were just on the margin of the town, while the third group were in the town ; these latter waters were all obviously polluted, while the fourth and fifth groups of wells in the heart of the town, and close to the brook, were all grossly polluted.

Waters from the Bunter Sandstone In the Urban District of A.

I.—Unpolluted Water above Town.

					PARTS PER 100,000.			
				Total Solids.	Free Ammonia.	Organic Ammonia.	Chlorine.	Nitrogen as Nitrates.
1.	Well, The Workhouse, No. 1.			40·0	·000	·005	1·45	·1
2.	„ The Mount	...	...	85·0	·001	·007	1·2	·1
8.	Spring at the Hall	...	...	85·2	·001	·007	1·4	Nil

II.—Passable Water ? (Surroundings of Wells to be Examined).

Near to the Town.

4.	...	...	...	...	...	56·0	·001	·004	2·5	·45
5.	...	...	...	...	...	51·0	·001	·009	2·5	·5

III.—Suspicious Waters, containing a Considerable Amount of Organic Matter which is Oxidized, but which may become Dangerous at any Moment.

Nearer to the Town (Total Solids, Chlorine and Nitrates higher).

6.	...	...	...	...	...	64·0	·001	·005	2·9	1·2
7.	...	...	...	...	...	52·0	·000	·002	5·2	1·0
8.	...	...	...	...	...	40·8	·003	·004	3·0	1·1
9.	...	...	...	...	...	72·8	·001	·005	2·5	1·0

IV.—DANGEROUS WATERS, CONTAINING AN EXCESSIVE AMOUNT OF ORGANIC MATTER OF ANIMAL ORIGIN WHICH AT ANY MOMENT MAY BECOME INJURIOUS TO HEALTH.

In the Town itself (Total Solids, Chlorine and Nitrates higher still).

					Total Solids.	Free Ammonia.	Organic Ammonic.	Chlorine.	Nitrogen as Nitrates.
10.	...	...	...	...	84·4	·001	·006	6·5	2·0
11.	...	...	...	...	95·0	·002	·012	9·5	2·2
12.	...	...	...	...	60·0	·001	·0105	8·2	1·8
13.	...	...	...	...	88·0	·000	·008	13·2	2·5
14.	...	...	...	...	86·0	·000	·009	16·0	2·3
15.	...	...	...	...	88·0	·001	·007	10·0	2·0
16.	...	...	...	...	110·0	·008	·012	10·5	2·1
17.	...	...	...	...	81·0	·001	·007	8·5	1·9
18.	...	...	...	...	75·0	·001	·006	8·5	1·8
19.	...	...	...	...	116·0	·001	·011	10·5	2·0
20.	...	...	...	...	80·8	·002	·011	5·5	1·9

V.—WATERS GROSSLY POLLUTED WITH ORGANIC MATTER OF ANIMAL ORIGIN.

Wells situated in the Lower Part of the Town near to the Brook.

21.	...	...	...	...	139·0	·002	·015	13·5	2·3
22.	...	...	...	...	80·0	·005	·018	9·5	2·8
23.	...	...	...	...	100·0	·002	·018	12·2	2·3
24.	...	...	...	...	77·0	·002	·016	8·3	2·1
25.	...	...	...	...	108·0	·002	·018	11·5	2·6
26.	...	...	...	...	96·0	·003	·016	10·0	2·1
27.	...	...	...	...	81·0	·12	·024	7·8	2·0
28.	...	...	...	...	96·0	·108	·012	11·2	2·4
29.	...	...	...	...	92·0	·015	·010	10·9	2·2
30.	...	...	...	...	80·0	·140	·013	10·4	1·9

CHAPTER VI.

On the Bacterioscopic Examination of Drinking Waters.

43. It is frequently of the greatest use to be able to estimate the number of bacteria per cubic centimetre, particularly with the view of testing the efficiency of the method of filtration in the case of river waters.

A thorough bacteriological examination of a water can only be made by a skilled expert and in a properly equipped bacteriological laboratory, but to count the number of colonies which can be cultivated from 1 c.c. of water is a comparatively simple matter, and it is well worth the while of the medical officer of health procuring the necessary apparatus for this ordinary bacterioscopic examination of waters.

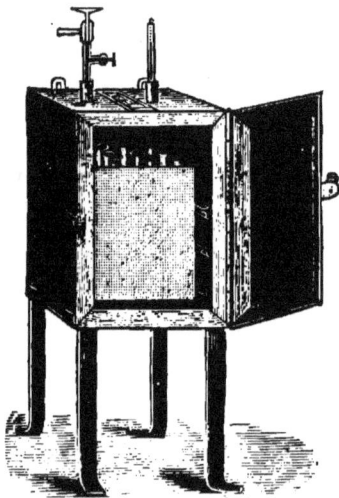

FIG. 18.

44. The appliances which are necessary consist of a steam sterilizer for the preparation of the culture media. A large ordinary potato-steamer will do very well for this purpose. The steamer should be covered with felt and Gooch's splinting to prevent waste of heat. The lid should also be covered with felt or other non-conductor, and should be provided with an outlet for the steam to insure its circulation *through* the steamer.

45. A small hot-air sterilizer (see Fig. 18) is required to sterilize dishes, flasks, bottles, pipettes, cotton-wool, corks, and other apparatus.

Two-ounce bottles with indiarubber bungs can be used for making

Fig. 19.

cultivations; they may be sterilized in the steamer, and the hot-air sterilizer may be dispensed with.

Forceps, glass tubing, platinum needles, and in an emergency

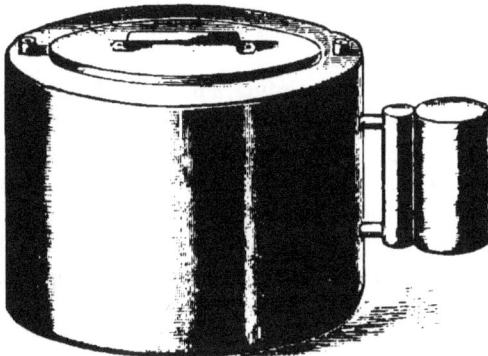

Fig. 20.

pipettes, may be sterilized by passing them through the flame of a Bunsen burner or a spirit-lamp.

46. An incubator set at 20° C. is also necessary. (See Fig. 19.)

The simplest form of a really useful apparatus consists of a water-jacket with a movable lid. (See Fig. 20.) The most reliable thermostats are those actuated by a capsule containing ether.

47. Glass dishes (Petri's dishes; see Fig. 21) with lids are extremely useful for making plate cultivations, although excellent work can be done with ordinary flat, clear glass bottles, or even by roll cultures in large test-tubes, or better with the small bottles which are used for samples of spirits or oil.

FIG. 21.

48. To measure the quantity of water to be taken, several 1 c.c. pipettes should be obtained; those made of thermometer tubing, and graduated in tenths of a cubic centimetre, are the most convenient.

49. As a culture medium for estimating the number of bacteria, 10 per cent. gelatine peptone is of the most general use.

To make it, take a small tin of Brand's extract of beef (beef jelly), and dissolve in 500 c.c. of water; add 5 grammes of peptone, which should be first mixed into a fine paste with a few cubic centimetres of the liquefied beef jelly; then add 2·5 grammes of common salt. Place the broth so made in a beaker or jar, and add 50 grammes of the best gelatine; gradually warm up by placing the beaker covered over with a piece of leadfoil in the steamer, and stir occasionally until the gelatine is dissolved, render neutral or *faintly* alkaline with carbonate of soda; finally steam for one hour. If there is any suspended matter, allow it to subside, and pour off the clear gelatine, or filter, and again sterilize.

50. In order to store the gelatine, a number of test-tubes or glass bottles should first be sterilized in the hot-air sterilizer at about 120° C. A bundle of cotton-wool and some good corks to fit the bottles should also be sterilized at the same time, the wool being required to plug the test-tubes with. The bottles may be plugged better by perforated indiarubber bungs, which should be previously steamed to sterilize them. They should have through the perforations pieces of glass tubing drawn out to capillary points, so as to permit of being instantly closed by applying a flame.

In a laboratory in constant working order, the gelatine is not kept more than a few weeks; but usually a medical officer of health will only have occasion to make bacterioscopic examinations in an emergency, so that it is important to him that his material should be so stored that it will keep properly as long as possible.

The usual method of keeping gelatine in test-tubes plugged with cotton-wool is open to two objections. First, the water in time evaporates, and the gelatine dries; secondly, moulds and bacteria may grow through the cotton-wool and contaminate the gelatine.

Gelatine may, however, be kept in bottles or stout glass tubes indefinitely if sealed hermetically at the moment of sterilization, the procedure being as follows :

51. The tubes, bottles, or two-ounce phials, the corks, cotton-wool, and perforated bungs being first carefully sterilized, 5 c.c. of the sterile gelatine is run into each. The method of doing this is to fix a funnel in a retort stand and to attach to its lower end a piece of indiarubber tubing closed by a pinchcock, and in the lower end of the tubing a piece of glass tubing drawn out to a point. The whole of this apparatus should be first sterilized ; then, if the gelatine is poured into the funnel, 5 c.c. may be easily introduced into each tube or phial without smearing its sides.

If the tubes or phials are for immediate use, they may be plugged by twisting plugs of sterilized cotton-wool into their necks; if it is desired to keep them longer, on the top of the cotton-wool insert a sterilized cork lightly, so as to permit the escape of air in the subsequent sterilization, or without the cotton-wool plug by means of a perforated bung.

The tubes are then to be placed in a wire cage in the steam sterilizer ; small bottles have the advantage that they can stand upright without a wire stand. They should then be steamed for half an hour on two separate days, and as soon as they have become sufficiently cool to permit of their being handled, they are to be taken out, the capillary ends of the glass tube are to be sealed, or the corks pushed well home, and when the gelatine has set the necks of the bottles may be dipped in melted paraffin. In the case of tubes plugged with cotton-wool, a small indiarubber cap should be placed over each.

52. To ascertain the number of bacteria in a water, the sample should be collected in a sterile strong glass-stoppered bottle, which should only be partly filled, to permit of the expansion of the water if it is frozen, for it is best to pack the bottle in a tin can surrounded by a freezing mixture of ice and salt in order to convey it to the laboratory. The necessity for this will be apparent from the following example of the manner in which micro-organisms multiply at ordinary temperatures :

ORGANISMS PER CUBIC CENTIMETRE.

	Day of Collection.	After standing 24 hours at 60° F.	After standing 48 hours at 60° F.
Derby tap water ...	16	59	Over 1,000

53. Having brought the sample to the laboratory, the bottle should be carefully washed on the outside with good tap water, and then be wiped dry with a clean sterile cloth.

The gelatine in four of the bottles should be just melted at about 90° F.; to one, $\frac{1}{10}$ c.c. of the water should be added; to a second, $\frac{1}{4}$ c.c.; to the third, $\frac{1}{2}$ c.c., and 1 c.c. to the fourth. The water should be added by means of the 1 c.c. pipettes, divided into tenths of a cubic centimetre. In adding the water, the gelatine should be exposed to the air as little as possible ; this is done safest, in the case of the bottles plugged with rubber bungs, by withdrawing the glass tubing from the indiarubber bung. The tube or bottle should again be closely plugged, the cotton-wool being ignited as it is plugged, and the gelatine and the water introduced should be intimately mixed by turning the tube round and round gently so as not to enclose any air-bubbles ; the gelatine is then poured into a Petri dish. If it is desired, the cultivation may be made in the bottle ; when this is contemplated, either 4-ounce flat bottles should be used or round ones ; in either case they must be of good white glass and free from air-bubbles. In the former case the bottle is laid on its side and the gelatine allowed to set in a film on one side ; in the latter case the bottle, with the cork or bung well fixed in, is plunged below *iced water*, while it is turned round evenly so as to spread the gelatine out in a uniform thin film, making a roll cultivation.

If it is desired to identify any of the colonies, the cultivation should be made in a Petri dish, but for the simple enumeration of colonies roll cultivations in clear glass round bottles (about 4 ounces capacity), plugged with perforated indiarubber bungs, will be found the easiest method. To count the bacteria, two small indiarubber rings should be slipped over the bottle about half an inch apart ; the colonies between the two rings being then counted, the rings should be brought together, and the top ring should then be moved half an inch further up, the colonies in that half-inch being counted, and so on until the whole surface has been gone over. It will facilitate the counting to have a line scratched with a diamond the whole length of the bottle.

54. The following table gives the number of organisms in various waters :

	Organisms per c.c.
Exceedingly pure waters	Less than 10
Pure waters	10 to 100
Suspicious waters	100 to 1,000
Impure waters	1,000 and upwards.

The standard of 100 organisms per cubic centimetre, which Koch suggests as the maximum for a drinking water, is quite as unreasonable as the old discarded standards of chemical purity.

The proper use of the method is purely comparative, and samples should be collected at the same time of waters from an unimpeachable source of a similar kind to the sample in question, and the results compared.

DETECTION OF SURFACE POLLUTIONS.

55. As the *Bacillus coli* is often found in water free from sewage pollution, Klein recommends search for the *Bacillus enteritidis*. Thresh gives the following simple method of detecting this latter organism :

To detect the B. enteritidis.—Place 10 c.c. of milk in a test-tube and sterilize at 100° C. for half an hour. Cool down to 80° C. and add 1 c.c. of the water to be examined. Keep at 80° C. for fifteen minutes. Then incubate anaerobically at 37° C. This can be done effectively by adding about 1 gramme of vaseline to the tube after mixing the water with the milk. When exposed to the temperature of 80° C. the vaseline melts and floats on the surface. After cooling the vaseline solidifies and prevents access of air.

Thresh states that " If the *B. enteritidis* be present, in twenty-four to thirty-six hours the casein will be precipitated and torn into irregular masses, which are frequently forced up against the cotton-wool plug, owing to the copious development of gas which occurs. The contents of the tube has a strong odour of butyric acid, and a strongly acid reaction."

INDEX.

Rebman Limited, 129, Shaftesbury Avenue, W.C.

www.ingramcontent.com/pod-product-compliance
Lightning Source LLC
Chambersburg PA
CBHW031753090426
42739CB00008B/990